# Living with the Weather

# FOG, MIST AND SMOG

Andrew Dunn

RSVP

RAINTREE
STECK-VAUGHN
PUBLISHERS
The Steck-Vaughn Company

Austin, Texas

*Living With the Weather*
# FOG, MIST AND SMOG

*Other titles:*
# HEAT AND DROUGHT
# RAIN, WIND AND STORM
# SNOW AND ICE

Published by Raintree Steck-Vaughn Publishers, an imprint of Steck-Vaughn Company

Printed in Italy. Bound in the United States.
1 2 3 4 5 6 7 8 9 0 02 01 00 99 98

**Library of Congress Cataloging-in-Publication Data**
Dunn, Andrew.
Fog, Mist, and Smog / Andrew Dunn.
    p.    cm.—(Living with the weather)
    Includes bibliographical references and index.
    Summary: Discusses fog, mist, and smog and the conditions which create them.
    ISBN 0-8172-5053-0
    1. Fog—Juvenile literature.
    2. Smog—Juvenile literature
    [1. Fog.  2. Smog.]
    I. Title.  II. Series.
QC929.F7D86   1998
551.57'5—dc21          97-22354

# CONTENTS

# OUT OF THE FOG

"In mist and cloudy sky one sees absolutely nothing. One fumbles with the feet on the uneven snow floor as if it were pitch-dark night." These are the words of Reinhold Messner, the first man to climb Mt. Everest without oxygen, speaking about his Antarctic expedition. It may seem strange that he should consider mist to be a danger, but many adventurers like Messner have found mist and fog to be treacherous, causing them to lose their way.

## Low clouds

The clouds in the sky, mist, and fog are all the same things. They are made up of moisture in the air. Mist and fog are just clouds at a low level. Smog, however, contains tiny solid particles. It is very unpleasant and unhealthy and is caused by machines, vehicles, and factories.

A cloud of smog hangs over the city of ▲ Seattle, Washington.

▶ A mist forms at dusk over the Arno River in Italy's ancient city of Florence.

# WHAT IS FOG?

Like clouds, fog and mist are made up of tiny droplets of water that hang in the air because they are too light to fall. But mist and fog do not float in the sky; they lie on or close to the ground or over the sea. They are often found in valleys and along coastlines.

## How far can you see?

When the air is clear and dry, visibility should be around 25 mi. (40 km), but the air is rarely clean enough to see that far. If visibility is poor but more than .5 mi. (1 km), weather experts say that it is misty. If the visibility is less than .5 mi. (1 km) then it is foggy.

▲ **Sun shines through the mist on an autumn day in the countryside.**

On a clear day you can see 25 mi. (40 km).

On a misty day you can see more than .5 mi. (1 km).

On a foggy day you can see less than .5 mi. (1 km).

## Smoke and fog

Smog is different from fog or mist. The word was invented in 1905 from the words "smoke" and "fog" because it was a kind of damp, foggy smoke. Smog has been a problem for centuries, ever since every home burned wood and then coal fires. The problem increased when factories began to burn large quantities of coal. Smog forms more quickly and lasts longer than fog, because it contains solid particles of smoke and dust.

▶ **You cannot see from one end of Mexico City to the other because of the smog.**

▼ **On a foggy day in London, England, visibility is reduced to under .5 mi. (1 km).**

## Chemicals and sunlight

There is another kind of smog, known as photochemical smog. It is formed when chemicals in the air react with bright sunlight to form solid particles that then hang in the air. The chemicals often come from car exhaust. Cities that tend to have still air, a lot of traffic, and a warm sunny climate, such as Athens and Los Angeles, suffer badly from this kind of smog.

## WHERE TO WATCH FOR FOG

Weather is not the same as climate. The weather is what is happening in a certain place at that moment in time. It might be sunny, snowy, rainy, or foggy. But the climate is the type of weather that is usual in a particular area. The climate of New York City, for example, is hot and humid in the summer, and cold and snowy in the winter. This is very different from the climate of Hong Kong, which has hot, wet summers and cool, dry winters.

### A likely climate

The climate in some places means that they hardly ever have foggy weather. But fog can be found in many different climates, from the tops of Hawaiian mountains in the tropical Pacific Ocean, to the cold fjords of Norway. Fog and mist can form wherever the air is moist, the sky is clear, and the wind is light. They are more likely to form in temperate countries rather than in the tropics.

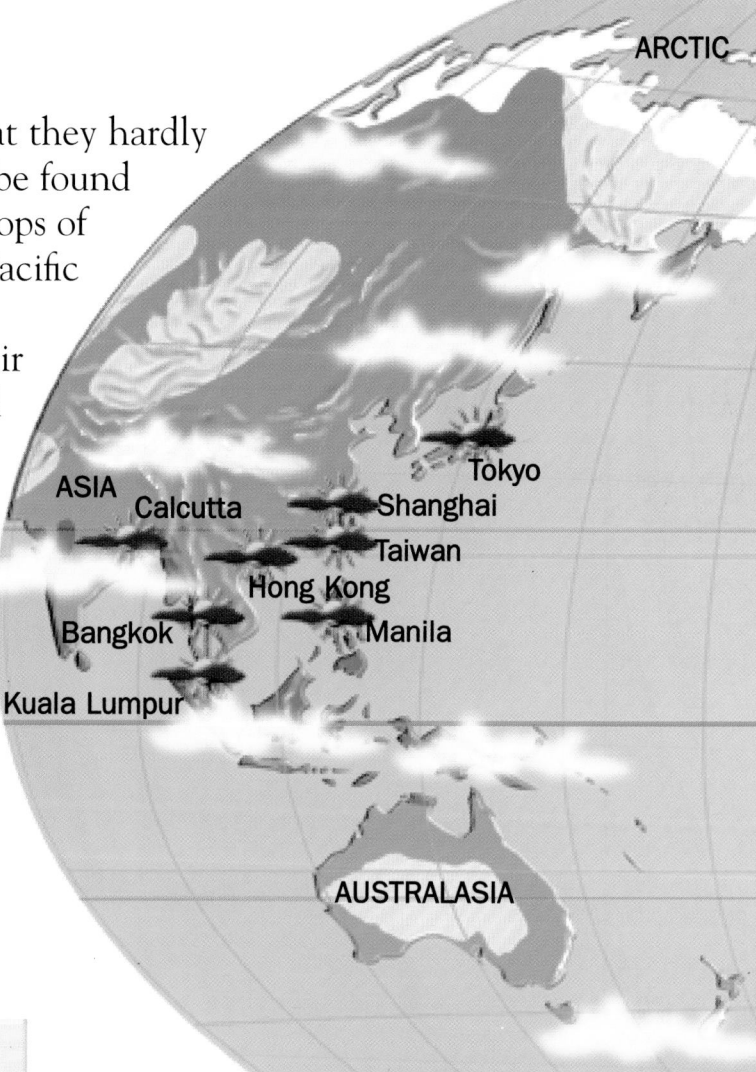

ARCTIC

ASIA
Calcutta
Tokyo
Shanghai
Taiwan
Hong Kong
Bangkok
Manila
Kuala Lumpur

AUSTRALASIA

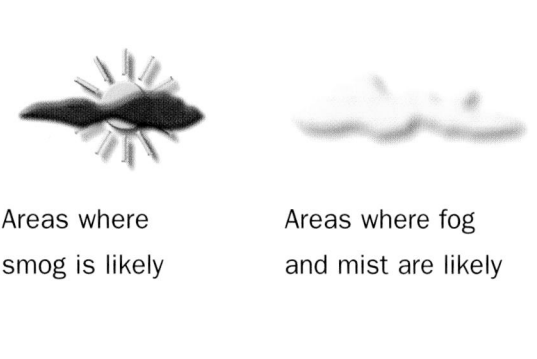

Areas where
smog is likely

Areas where fog
and mist are likely

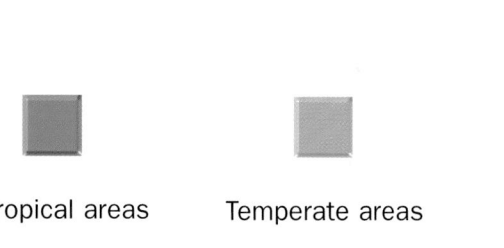

Tropical areas

Temperate areas

Arid areas

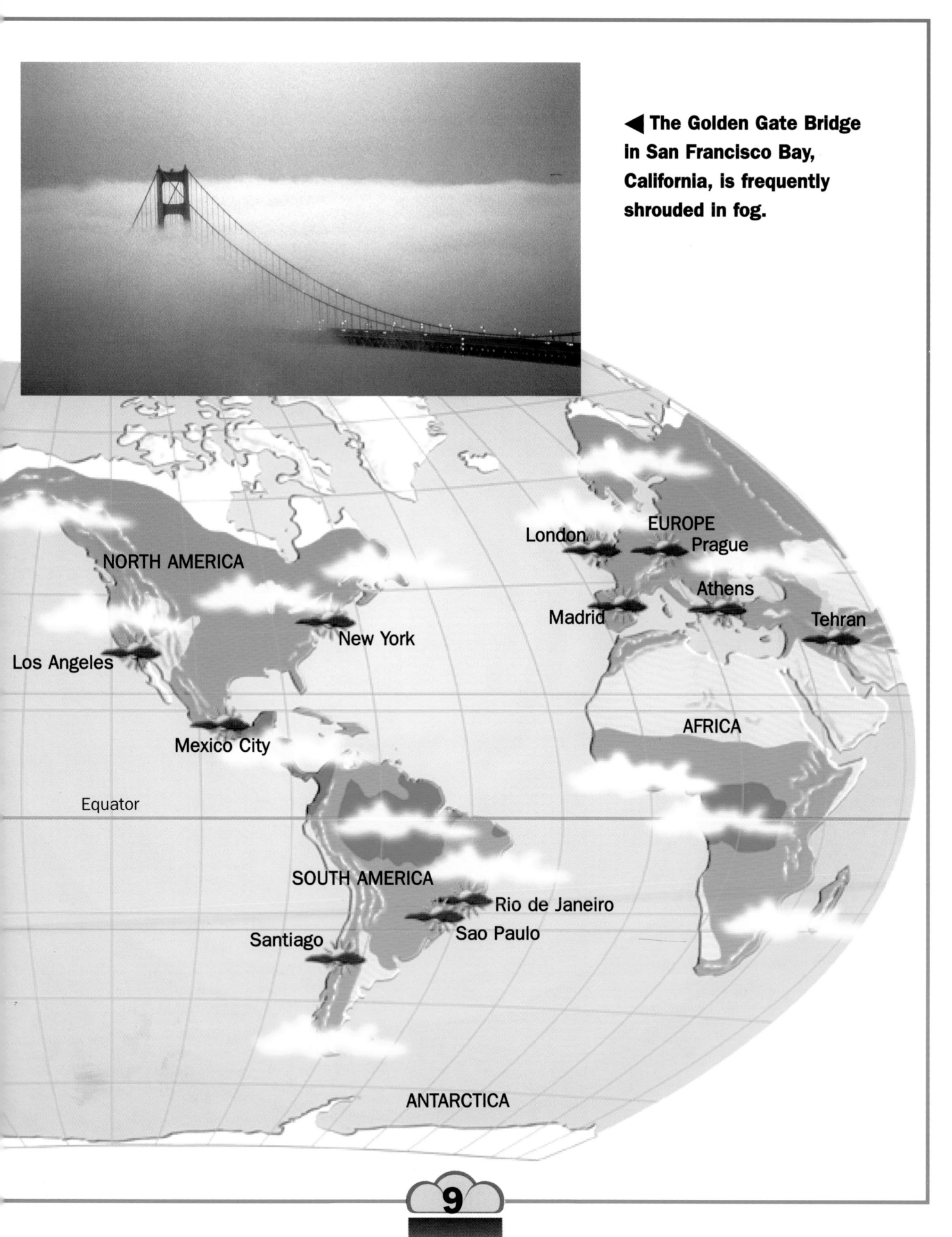

◀ The Golden Gate Bridge in San Francisco Bay, California, is frequently shrouded in fog.

NORTH AMERICA

Los Angeles

New York

Mexico City

Equator

EUROPE

London
Prague

Athens

Madrid

Tehran

AFRICA

SOUTH AMERICA

Rio de Janeiro

Sao Paulo

Santiago

ANTARCTICA

## WHERE TO FIND SMOG

Fog and mist mostly form in damp places: in mountain valleys, near rivers or marshlands, and along coastlines. But smog is much more likely to form in towns and cities than in open country. This is because in heavily populated areas there are many more of the particles of dirt and smoke that enable smog to form. Industrial pollution also makes smog of any kind more likely.

▲ A woman tries to block out dirty traffic fumes in Kuala Lumpur, Malaysia.

▶ Slow-moving traffic causes more pollution and more smog than fast-flowing traffic.

## Hot, dry smog

Photochemical smog can form even when the air is dry. It tends to be worse in large cities on hot sunny days.

Then, there are clear skies, thousands of cars, and sunlight to turn exhaust fumes into a thick haze that stings the eyes. Another major cause of photochemical smog is chemicals given off by plants, and pine trees in particular. You can sometimes see a smoggy haze over pine forests.

### HOW CLEAN IS YOUR AIR?

Line a funnel with a coffee filter, and place the funnel in the neck of a bottle. Then put the bottle outside, and leave it to fill with rainwater. Empty the bottle, and let the filter paper dry. Lay it out flat and, using a magnifying glass, inspect it for black specks. These are particles of soot.

◀ Factories pour out thousands of tons of pollutants into the air every year. Old factories tend to be worse polluters than modern ones.

# WHAT CAUSES WEATHER

The world's weather takes place in a thin layer of air surrounding the earth, called the atmosphere. Here, as air and water are heated by the sun, or become cooler, weather conditions such as wind, rain, snow, or fog are created.

## Water in the air

The air is full of water in the form of a gas called water vapor. In the earth's atmosphere there are about 3.4 sextillion gallons of water. Air cannot hold very much water vapor, however, and if there is too much, the vapor condenses.

This means that it forms droplets of liquid water. When the air is holding the maximum amount of water vapor it can, it is called "saturated." Once the air is saturated, the water vapor may form clouds, fog, mist, or dew. This is known as the "condensation point" or "dew point." Warm air can hold more vapor than cold air, so the condensation point depends on the temperature of the air.

▼ **The earth's atmosphere helps protect us from the harmful effects of the sun.**

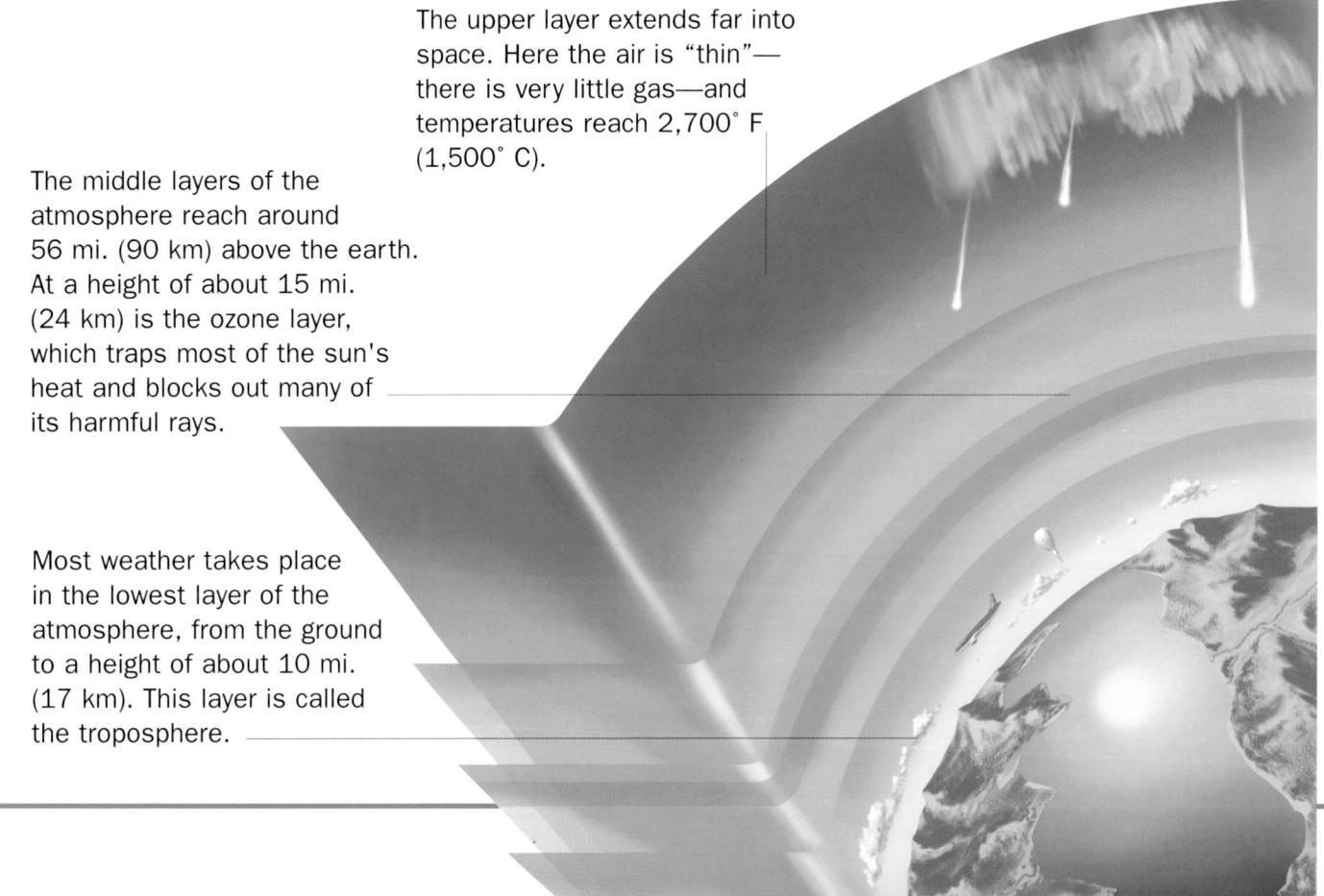

The upper layer extends far into space. Here the air is "thin"— there is very little gas—and temperatures reach 2,700° F (1,500° C).

The middle layers of the atmosphere reach around 56 mi. (90 km) above the earth. At a height of about 15 mi. (24 km) is the ozone layer, which traps most of the sun's heat and blocks out many of its harmful rays.

Most weather takes place in the lowest layer of the atmosphere, from the ground to a height of about 10 mi. (17 km). This layer is called the troposphere.

▲ **People crossing Westminster Bridge, London, in a thick fog**

## Cloud formations

There are different types of clouds because condensation can happen in several ways. When a large pocket of air slowly rises, it cools evenly. Water vapor condenses at the same height, so a flat sheet of cloud, called stratiform or stratus clouds, forms.

Sometimes the ground is warmed by the sun and gives off heat. The air above the ground rises in bubbles. Water vapor starts to condense at a certain height, but because the air is still rising, the clouds form high piles like balls of cotton. These are known as cumuliform or cumulus clouds.

Air moving over hills is forced to rise and condenses into orographic clouds, which means "mountain clouds."

## What are clouds?

Clouds are made up of floating water droplets, often mixed with tiny ice drops, and there are many different types. Clouds form when water vapor condenses or freezes out of the air to become water or ice.

## Masses of air

The lowest layer in the atmosphere is called the troposphere, and air tends to travel through the troposphere in pockets or masses, in which all the air has the same moisture and warmth. If some change makes the water vapor condense, this mass forms a cloud.

**CLOUD COVER**

★ Cloud droplets are a thousand times smaller than a raindrop.

★ Some spectacular clouds form around mountaintops. Table Mountain, just outside Capetown in South Africa, is often capped by a fine cloud, known locally as the Tablecloth.

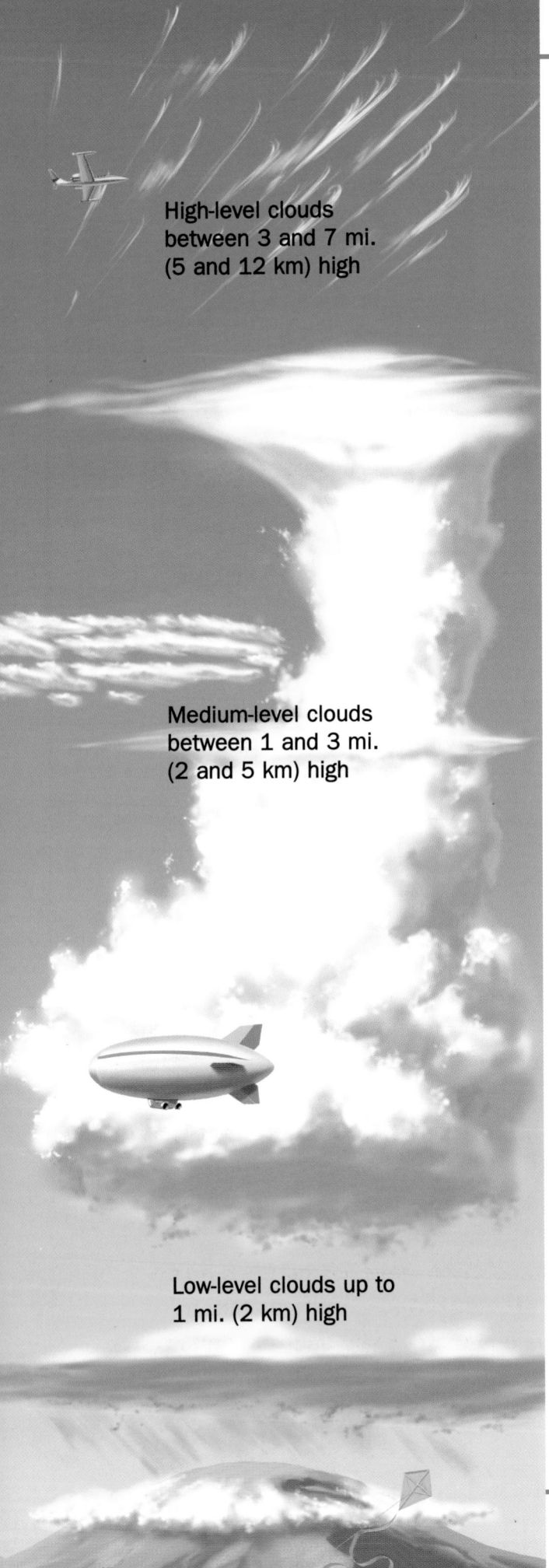

High-level clouds
between 3 and 7 mi.
(5 and 12 km) high

Medium-level clouds
between 1 and 3 mi.
(2 and 5 km) high

Low-level clouds up to
1 mi. (2 km) high

## TYPES OF CLOUDS

Clouds are the quickest guide to what weather is on the way. Meteorologists classify different types of clouds into families, and families into species—in just the same way that biologists classify plants and animals. To identify which family a cloud belongs to, you need to look at the height of its base and then its shape.

### Low-level clouds

Low-level clouds have a base between ground level (on the ground they are fog) and 1 mi. (2 km). They are the layered sheets of stratus, bumpy sheets called stratocumulus, and the fluffy piles of cumulus. Another type called cumulonimbus can have a low base but often reaches high into the sky.

### Medium-level clouds

Medium-level clouds are found between 1 and 3 mi. (2 km to 5 km). Here you find stratus clouds, such as altostratus and nimbostratus, and also altocumulus.

### High-level clouds

High-level clouds are mostly made of tiny ice crystals. Above 3 mi. (5 km) the air is quite dry, so any clouds tend to be thin, feathery clouds called cirrus, though thin sheets of cirrostratus and wispy heaps of cirrocumulus can also form at these heights—up to 7 mi. (12 km).

## MAKE YOUR OWN CLOUDS

There are two ways you can make clouds in a bottle. One way is to pour warm water into a plastic bottle until it is about 2 in. (5 cm) deep. Screw on the cap. Then squeeze the bottle hard to squash the air and then release it. This makes the air expand and contract, and as this happens a mist or cloud should form.

Another experiment you can try is cooling the air so that it is unable to hold all its water vapor, and some of it condenses. You can make this happen by filling a metal tray with ice and allowing it to get really cold. Next put 1 in. (2 to 3 cm) of warm water in a jar and place the cold metal tray on top of the jar. A cloud will form as the warm, moist air meets the cold air at the top of the jar.

## Rain clouds

Several types of clouds may produce rain. They usually have the word "nimbus" in their name. The most common rain cloud is the gray and threatening nimbostratus. Cumulonimbus can produce violent storms. Stratocumulus usually produces drizzle. It is often a sign of bad weather to come. Fluffy cumulus usually means nice weather, until it clumps together to form larger clouds. Orographic stratus clouds, formed as moist air is pushed up a hillside, often produce rain, too.

◀ Planes often leave behind them a thin strip of cloud called a contrail. This is because aircraft engine fumes contain water vapor. As the water vapor cools, it forms the contrail.

15

## HOW MIST AND FOG FORM

Mist and fog form in just the same way as clouds but at ground level. At night, without the heat of the sun, the ground cools. The ground in turn cools the air above it. The cooler air has to release some of its water vapor, and as a result droplets of water—dew—form on the grass or any other cold surface. But the temperature of the air is not the only factor that determines the dew point. If the air contains a lot of water vapor, dew will form at a relatively high temperature. This is because moist air only has to be cooled a little before it is forced to release some of the vapor it contains.

▲ When water vapor reaches dew point, droplets of water will form on a cold surface, such as this spider web.

## When and where?

There are different types of fog, but in general fog tends to form at the beginning of the day, because the air reaches its dew point after a long, cool night. In temperate climates this usually occurs in autumn and winter when the nights are long and cold. Fog also forms at night, especially if there are clear skies when the temperature drops fast. If there is only a thin layer of moist air, there will be a dew. If there is a deep layer of moist air, fog will form.

## Still and silent

Fog usually forms only when there is little wind. This also means that the fog lingers until a breeze stirs up and disperses the fog or until the sun comes out and warms the ground. When this happens, the fog sometimes begins to evaporate from underneath, so a layer of fog may be left floating off the ground. Although visibility may be good on the ground, you cannot see the sun. This is called fog stratus.

## Thick fog

Fresh air is not as clean as you might think. It is always filled with tiny, floating particles. These could be particles of salt from sea spray, desert sand, or pollution. There are usually five or six million of these particles in every quart (liter) of air over land, and one million at sea. These particles provide surfaces for water vapor to condense on. So the thickest fogs occur where there is bad pollution.

### CONDENSATION

Watch condensation form by half-filling a dry glass with ice cubes. Soon tiny water droplets will appear on the outside of the glass as water vapor condenses from the air cooled by the icy glass.

◀ If you wear glasses, you will have noticed that they steam up when you move from somewhere cold into a place where there is warm, moist air.

## TYPES OF FOG

Fog and mist form when air cools near the ground. This can happen in two ways, forming either radiation fog or advection fog.

## RADIATION FOG

The most important kind of fog is called radiation fog. It is the fog that forms over land as air cools, often on clear nights with a faint breeze. It is called radiation fog because it is formed as the ground gives off, or radiates, the heat it

▼ **Drivers can find they're in patchy radiation fog without warning.**

absorbed during the day. Radiation fog is common over damp areas such as ponds, streams, or marshes or ground that has recently had rain.

## Cloud cover

Fog is unlikely on cloudy nights, because the clouds reflect the radiated heat back to the ground. The air usually doesn't cool far enough to reach its dew point. Strong winds also keep radiation fog from forming, because a thick layer of air does not hang around long enough to cool. On the other hand, if there is no wind, only a very thin layer of air becomes cold enough to form fog. This can sometimes be seen early in the morning, when mist can hide the legs of cattle in a field, but not their bodies.

## Danger on the roads

Radiation fog is often patchy, which is especially dangerous for drivers. One moment the air is clear, and then suddenly it is difficult to see the road, let alone the car in front. If the driver ahead brakes suddenly, an accident can easily happen.

## ADVECTION FOG

Advection fog is not formed by the ground cooling, but by cold air moving into a moist area, or moist air reaching a cold area. It can form even in strong winds and will not disperse until the air has been warmed by the sun.

▲ **Moist air turns into fog over the cool surface of the Loire River, in France.**

How can you tell what kind of fog you're in? Radiation fog is very still, while advection fog flows. Another difference is that advection fog usually forms during the day, while radiation fog forms at night.

## FRONTAL FOG

Frontal fog is different still. It forms at the base of the front of a warm mass of air, known as a warm front. Rain falls where the warm air meets cold air, and the warm air can become very moist.

## Valley and hill fog

There are certain times and places where fog is more likely than others. Fog and mist are made up of cooling air, which is heavier than warm air. Cold air tends to sink, and as a result fog and mist are more likely in valleys than on hilltops. Mist is particularly likely in river valleys, where the air is damp and cool. Hill fog is really low clouds. It forms when air cools as it rises to pass over high ground.

▲ **Because cold air sinks, mist and fog are often found in valleys.**

## Sea fog

Sea fog can form as warm air moves over cold sea. This often happens near Newfoundland, Canada, where warm water and air meet the cold water coming south from the Arctic.

## Fatal fog banks

There is a spot south of Newfoundland, in Canada, where winds warmed by the ocean currents of the Gulf Stream in the south meet cold currents from the north. There are often thick banks of fog. To add to the danger, in the same area there are icebergs from Greenland. It was there that the liner *Titanic* hit an iceberg in 1912 and sank, drowning 1,500 people.

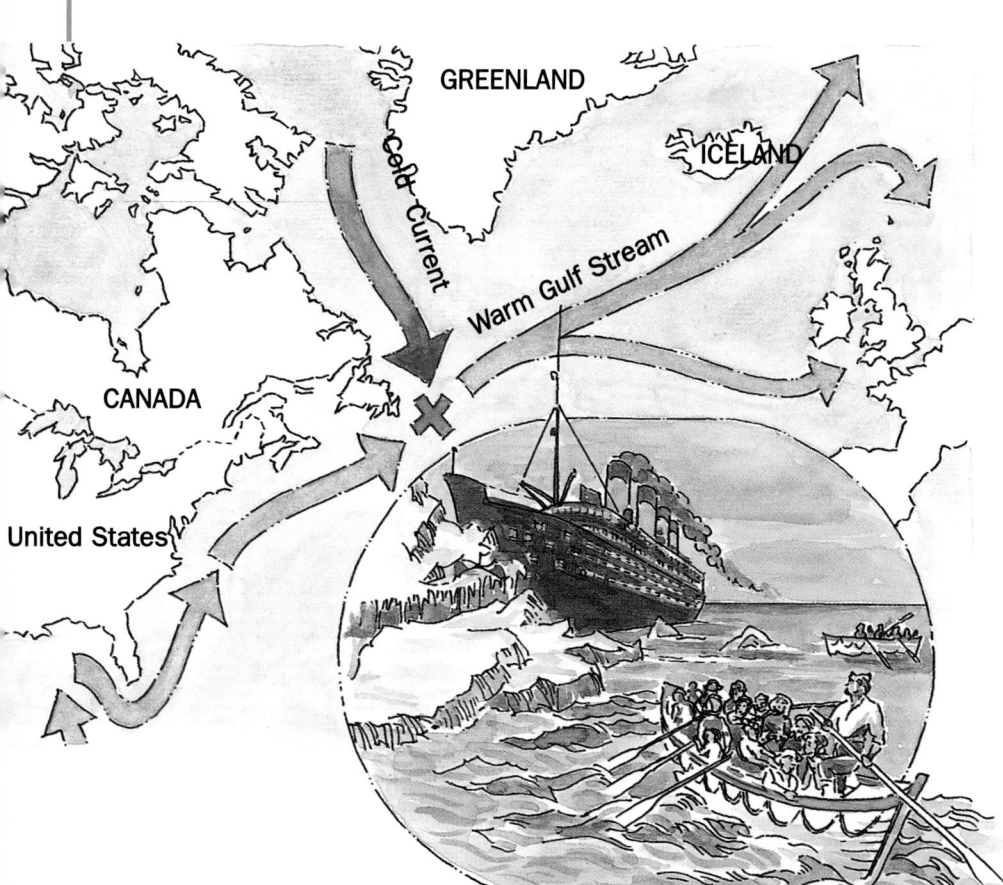

◄ **The coast of Newfoundland in Canada is particularly dangerous for ships, as fog and icebergs are common.**

◀ **Mist forming in the canopy of a tropical rain forest, as the warm, wet air cools and releases its water vapor.**

▼ **Steam fog in Arctic waters, formed as cold air from the open ocean mixes with warmer coastal air.**

## Steam fog

In Arctic regions, steam fog can form when cold air blows over warmer water. This forms in the same way as steam from a kettle. A thin layer of air that has been warmed by the water mixes with the cold air and cools. The same sometimes happens over a warm lake on a cold night.

## Rain forest fog

Mist forms in the rain forests of the world because the air is hot during the day. It can hold a lot of water vapor, which comes from the trees and plants. As the air cools in the evening, the valleys fill with mist.

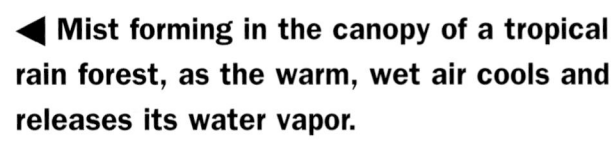

**IN OR OUT?**
★ On Gomera, one of the Canary Islands off northwest Africa, there is a small rain forest on a mountain peak with dry slopes below. As a result local patches of fog produce a strange phenomenon: sometimes it is possible to stand with one arm in fog and the other in sunshine.

▲ Hoarfrost left behind by freezing
fog transforms everything it touches.

## Mountain mist

If peaks jut into a moist stream of air, mountain mist can form even when it is not cold. The air cools as it rises over the mountains, forming mist or fog. This happens on Pacific islands such as Tahiti and Hawaii. Their mountaintops are often shrouded in mist.

▲ **Mountain peaks in Nepal rise out of clouds of early morning mist.**

### DISASTER AT TENERIFE

★ In March 1977, at Los Rodeos Airport in Tenerife, fog helped cause one of the worst disasters in the history of flying. Two jumbo jets were waiting to take off, as an afternoon fog rolled down onto the runway. The first plane drove on to the runway to take off. The second went to the other end of the runway. Both pilots misunderstood their instructions and began to take off through the fog. The planes collided at high speed. Loss of life stood at 583, although 69 people escaped.

## Freezing fog

Sometimes air temperature can be below freezing, yet the air is still moist. This can produce a fog of supercooled droplets of liquid water. They can stay liquid even below freezing point, as cold as -4° F (-20° C), but will freeze instantly on touching any surface. This type of freezing fog coats everything—tree branches and plants, lampposts, even men's beards—with a white coating of what is called hoarfrost. It is very slippery for people and cars. Another problem is that it builds up on electricity towers and TV transmitters, sometimes causing them to collapse under the weight.

## WHAT CAUSES SMOG?

Tiny particles of dust in the air help fog and mist form. Smoke does the same, because the tiny soot particles give the water vapor somewhere to condense. Smog is a deadly combination of smoke and fog. When pollution is in the air, smog forms more easily than ordinary fog and lasts longer.

### Trapped smog

Air is normally warmest near the ground and gets colder higher up. But sometimes the weather turns upside down. In autumn and winter there are often temperature inversions, where a layer of warm air traps and holds a cold layer beneath it. Pollution in the air cannot escape upward. Temperature inversions can happen among mountains as well, trapping hill fog in the valleys. In cities, smog can form in the cold air at ground level and stay for days. Unfortunately some cities are located in areas where temperature inversions are more likely to happen because they tend to have low rainfall and light winds. Mexico City is one such city. Although it produces similar levels of pollution to other cities, the local weather conditions result in greater pollution problems.

▼ A layer of warm air can trap a layer of cold air beneath it. This is called a temperature inversion and means that pollution cannot escape and may stay trapped for several days.

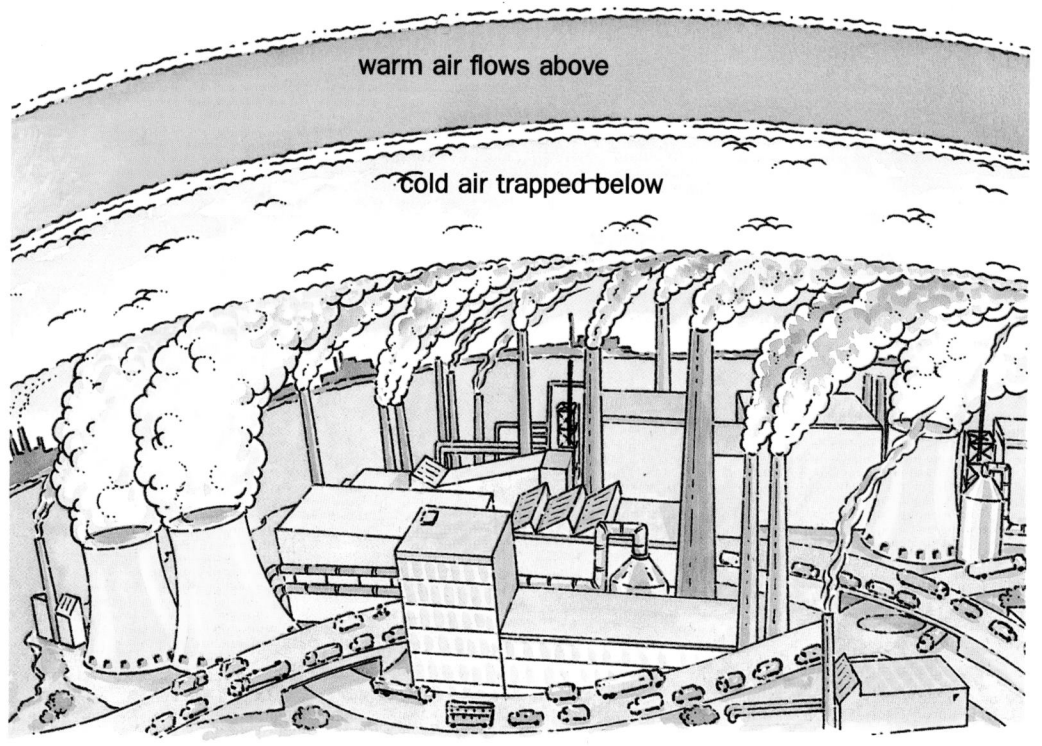

warm air flows above

Cold air trapped below

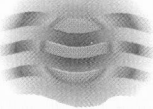

## FIREWORK SMOG

★ Smog may occur on days when there are many firework displays, such as November 5 in Great Britain or July 4 in the United States. If there is a temperature inversion, the amount of smoke from the fireworks builds, and as the air cools the pollution sinks to the ground, forming a thick smog.

▼ **Levels of pollution are so high in many cities today that cyclists wear face masks to filter the dangerous particles from the air.**

## Action against smog

Smog can be reduced by cutting down the levels of pollution. As a direct result of the terrible 1952 London smog and the public outcry, the British government began to take action. Smoky fuels were banned from many areas, "smoke-free zones" were created, and factory owners were forced to cut down smoke from their chimneys. The worst smogs became a thing of the past in London. Yet factories still pour out pollution of other kinds, and this old-fashioned smog still forms in many cities.

## PHOTOCHEMICAL SMOG

Photochemical smogs are not really smogs, because they are not a kind of fog. The chemicals that produce this smog are mainly sulfur dioxide and carbon dioxide from power plants and factories, and carbon monoxide and nitrogen oxides from the exhaust fumes of cars and trucks.

### Lethal gases

When sunlight falls on these invisible gases in still air under clear skies, they turn into a mixture of lethal gases, including a form of oxygen called ozone. Ozone is vital in the upper atmosphere, where it absorbs the harmful rays of the sun, but on the ground it is poisonous to breathe.

### Harmful effects

Together these chemicals form a visible smog, which is irritating to the eyes and lungs and can be dangerous to children and adults with asthma or other lung conditions. The smog may also spread out from cities to the countryside, where it damages crops.

### Trapped air

Temperature inversions play their part in forming photochemical smogs. In coastal cities such as Los Angeles, cold air blowing off the ocean can be trapped under warm air above, and when smog forms it cannot move.

▼ **Skyscrapers in Los Angeles rise through a thick layer of smog.**

## Natural smog

Photochemical smog can form naturally. Plants produce it—particularly pine trees. They give off highly scented waxy chemicals, and sunlight reacts with these to form a bluish hazy smog. Some pine-covered mountains have been named after this "natural smog," including the Blue Ridge Mountains of Virginia, and the Blue Mountains in Australia.

◄ Chemicals mix with moisture in the air to form acids that eat into stonework, causing it to crumble away.

# FORECASTING FOG

▲ A low-lying mist hangs over a road, in an early-morning frost.

Modern weather forecasting for a day or two ahead is much more accurate than it used to be. But fog and mist are very difficult to forecast precisely. The most a weather forecaster can say is whether fog is likely in a particular area. This is because if the conditions needed for fog to form vary even slightly, it will not form. If the air is drier than expected or if the temperature does not drop as much as forecast, the dew point will not be reached and water vapor will not condense. If a wind picks up, fog will not form. These conditions can vary over short distances.

## LOCAL GEOGRAPHY

Knowing your local geography helps you figure out your local risk. On days when fog is likely, it tends to form in low-lying river valleys rather than on exposed hillsides. Fog is more likely if the ground is damp from recent rain than if the weather has been dry.

## CLEANING THE AIR

It is easier to predict when fog will clear. Fog clears in two different ways. Either wind blows it away, or the sun "burns" it off from above. In winter, when the sun is low in the sky all day and its rays are weak, if there is little wind, fog may not clear for several days.

▲ **Weather forecasting relies on local weather stations for detailed information.**

▲ **An officer measuring rainwater at a weather station**

### A VISIBILITY DIARY

Use a map and ruler to calculate the distances to landmarks you can see from your home—hills, tall buildings, trees. Then keep a diary of how far you can see each morning by noting how far away the farthest clear landmark is.

## AIR QUALITY

In many American cities, local weather bulletins give forecasts of the air quality for the coming day. These forecasts predict the chances of the air being smoggy or polluted. This information allows people with breathing difficulties to prepare themselves if the pollution is going to be bad.

# LIVING WITH FOG

◀ Lighthouses sound horns in the fog, because their lights may not be seen.

# TRAVELING IN FOG

Fog can be alarmingly confusing. It can also be very dangerous for travelers in ships or aircraft, driving on the roads, or even out walking.

## Fog at sea

At sea, even though most ships now have radar that can "see" through fog, ships still collide. Lights may not penetrate more than a short distance, and big ships cannot stop quickly. If they are on a collision course, they may not see each other until it is too late. There is also the risk of hitting rocks or icebergs.

## On the map

In fog, it is particularly important for sailors to be sure of their exact position. Near land, they take points from the ship's radar and compare them with maps called charts. Out in midocean, where it is less crowded, they navigate by satellite "fixes." But all ships must have human lookouts keeping an eye on the surrounding water as well as radar.

## NAVIGATION

★ Most modern ships navigate by using electronic systems as well as by compass. Many use radio signals to find their position. The signals are beamed from points on land or in space (from satellites). Ships' computers pick up the radio beams and figure out where the ship is positioned in relation to the beam.

## Foghorns

Lighthouses warn passing ships of dangerous waters, but in fog they may not be seen. So most lighthouses sound deep, booming foghorns. Sea fog can play tricks with sound, however, and it can be difficult to be sure exactly where the noise is coming from. Ships, too, must carry sirens to warn each other of their presence in the fog.

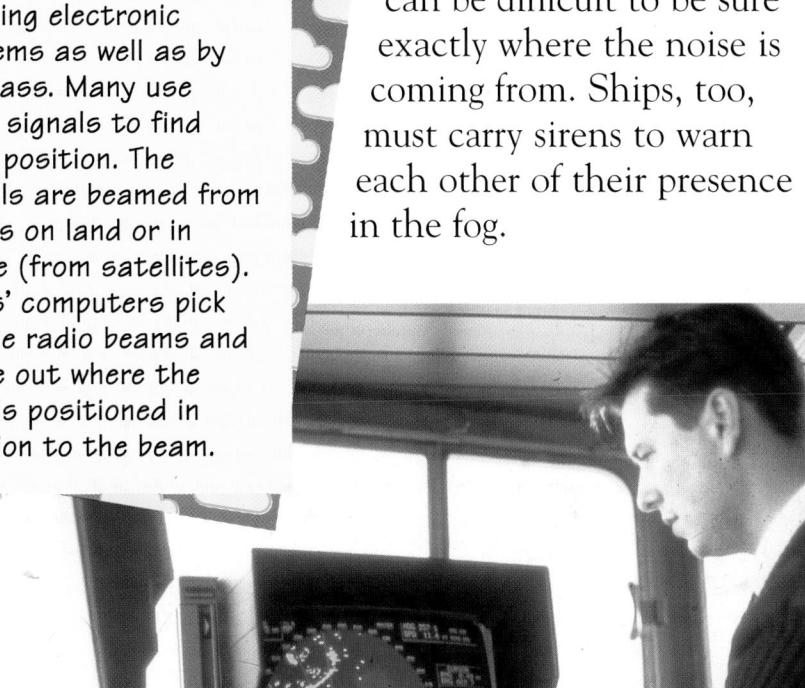

▲ A radar screen shows the position of the ship and the movements of nearby vessels.

## FLYING IN FOG

Air travel can be badly disrupted by fog, despite all the modern technology that aircraft carry. Usually though, it is not directly because of the fog. Modern aircraft are increasingly able to operate in bad weather and low visibility. Pilots have to be able to take off, fly, and land by instruments alone, without being able to see the ground. Pilots in the latest, sophisticated aircraft know where they are, which way they are going, and how near other aircraft they are. They know this from their equipment and landing guidance signals from the ground, which guide planes precisely onto the runway.

▼ **The most dangerous time for planes in the fog is when they are on the ground at the airport. In the air, air traffic controllers and computers can guide them.**

## Fog alert

Nevertheless, flights are sometimes diverted to land at other airports, and takeoffs are delayed. This is for two reasons. The first is that in poor weather and low visibility, the air traffic controllers increase the separation between aircraft. That means fewer can land or take off in any given time, so the airport cannot handle as many flights. Some have to go elsewhere to land or wait longer to take off.

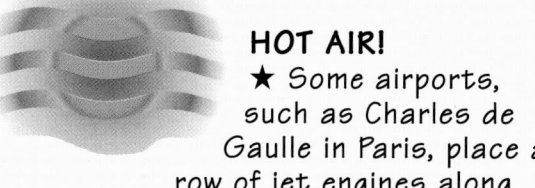

**HOT AIR!**
★ Some airports, such as Charles de Gaulle in Paris, place a row of jet engines along the side of the runway. When fog appears, the engines are run, to create wind and heat to clear the air. The British tried a similar trick during World War II, when flamethrowers were used to clear fog for a few minutes so that returning bombers could land. It was reported some pilots thought the flames were worse than the fog!

◄ Baggage handlers take the luggage off a plane that is stranded in the fog. The passengers will have to wait until the fog lifts or make other arrangements for their journey.

Flights are also stopped because not all aircraft have the latest equipment. Many major airports sometimes become fogbound in winter. They have some of the most sophisticated guidance equipment in the world, but if an incoming aircraft does not have instruments that can understand the signals from the airport, the pilot may be unable to land safely.

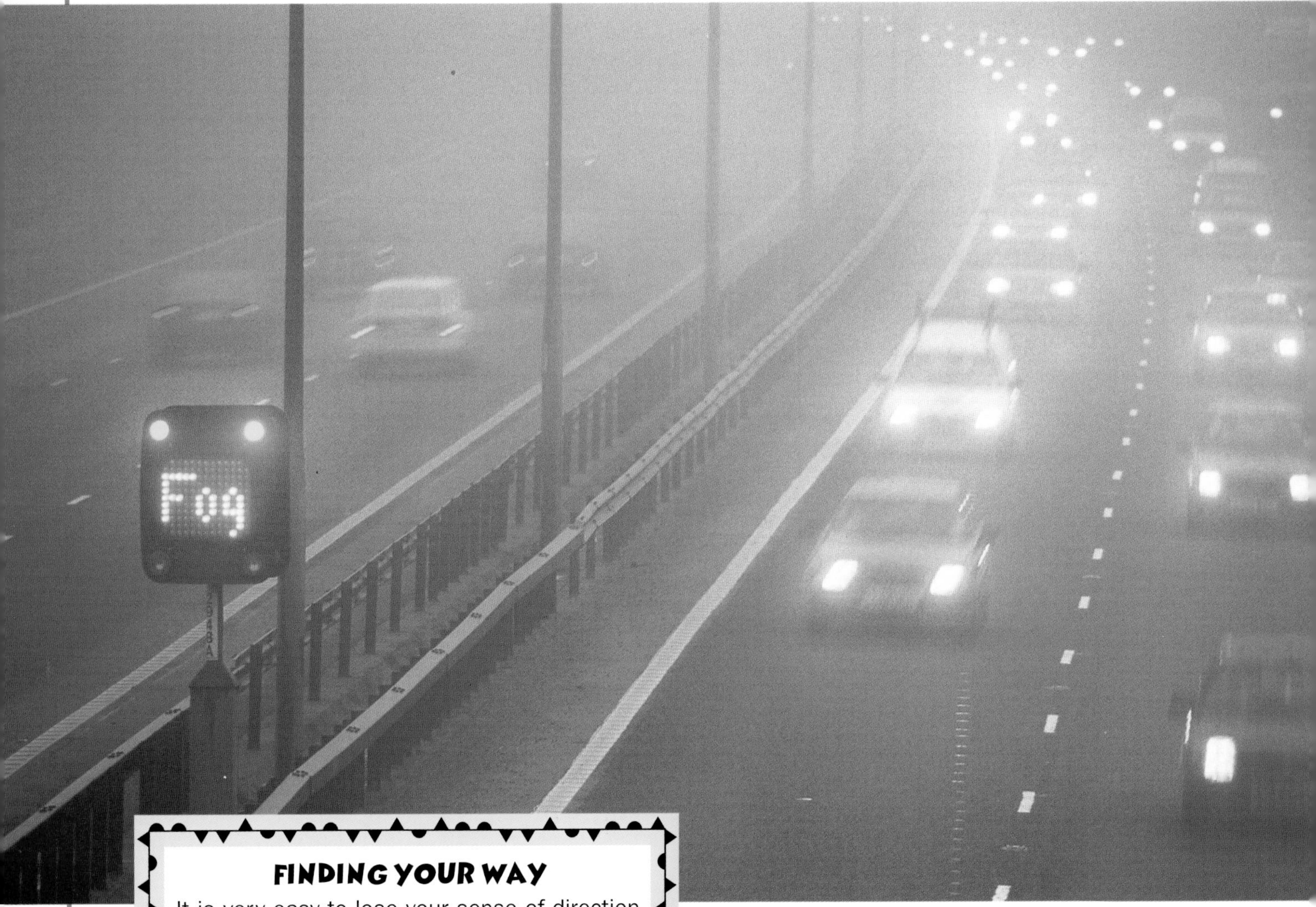


### FINDING YOUR WAY

It is very easy to lose your sense of direction when you can't see. Choose an object some distance away and get a friend to blindfold you with a scarf so that you cannot see. Aim to walk toward the object in a straight line—your friend can warn you if you are likely to trip over anything. When you think you are nearly there, stop and have a look. How well did you guess distance and direction?

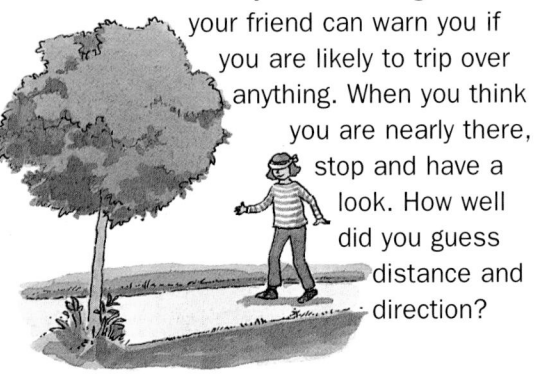

## FOG ON LAND

Driving in thick fog is almost impossible. Fog can be patchy, and drivers can go from bright sunshine into thick fog in a moment. Full headlights make things worse as a lot of light is reflected back by the water droplets. It seems as if a white screen has appeared in front of the driver. In foggy weather, drivers need to slow down. Then they have time to stop if there is an accident ahead.

◀ **In foggy conditions cars should slow down and put on their fog lights.**

▼ **Hill fog is a hazard for walkers. It is vital not to stray from marked paths or roads.**

Freezing fog can be more dangerous still. As well as making it difficult to see the road, it causes "black ice," invisible slippery ice on the road's surface.

## Fog on the roads

Even when fog is common and drivers know they have to take extra care, accidents due to fog often occur. In the mountains of Tennessee there are frequently dense fogs, which affect visibility. Drivers suddenly find themselves unable to see more than a few feet around them, and they are forced to brake suddenly. The driver behind doesn't see the driver in front trying to slow down, and the cars crunch together. Each year there are crashes involving hundreds of cars and many people lose their lives in such freakish accidents.

## Lost on foot

Walking or groping through city streets in thick fog is bad enough. Walking on hills or moors in fog is an eerie experience. With no road to follow, it is easy to get lost. It is essential to have a map and compass and to know how to use them. When landmarks disappear, it may be necessary to check your position with a compass every few feet. People lost in fog sometimes walk around in circles, convinced they are walking in a straight line toward safety.

## FOG AND NATURE

Fog can sometimes be useful. When a winter night is cloudy, it is unlikely to be as cold and frosty as it would be if the skies were clear. The same is true of fog: it acts like a blanket, protecting the ground from losing any more heat, preventing frost unless the air is already very cold.

## Protecting crops

Commercial crop growers are quick to protect their crops from frost. Some use large fans to keep breezes blowing over their fields, but these fans are expensive to run.

### CATCHING DEW

On the Spanish island of Tenerife, it hardly ever rains. But the farmers there have found a way to use fog and dew to water their crops. They plant their vines in hollow bowls dug in the hillsides. On cold misty nights, dew collects on the leaves and runs off them to water the thirsty vines.

▲ **California redwood trees collect moisture from sea fog on their needles. The pines may even feed off the nutritious minerals carried in mist and fog.**

In California, fruit trees are protected from frost by artificial fog. Overhead water pipes squirt water through very small holes, making a fine mist that can hang in the air for some time.

## Drinking fog

It may not rain for years in the Namib Desert in southwest Africa. The most important source of moisture is sea fog. The Welwitschia plant survives there because its big floppy leaves are able to soak up moisture from the fog.

▲ The Welwitschia plant sometimes looks as if it is dead, but then its leaves soak up moisture from sea fog and come back to life.

## Desert creatures

The tiny tenebrionid beetle drinks the Namib fog by climbing to the top of a sand ridge and facing backward to the incoming fog. The fog condenses on its back, like dew, and trickles down into its mouth. The Namib dune beetle digs little trenches in the sand, which collect moisture as the fog sweeps in. In Australia there is a lizard living in coastal deserts that doesn't need to drink at all. It simply absorbs foggy moisture through its skin.

## Trapping fog

In a village in the Andes Mountains of Chile, rainfall is so scarce that the people have learned to trap fog with rope nets strung out along the hillsides! Water condenses onto the nets and trickles down them into tanks below, collecting up to 15,800 gal. (60,000 l) a day.

▲ The Thorny Devil does not need to drink at all. It gets all the water it needs from moisture in the air.

## THE EFFECTS OF SMOG

The air in many cities has been polluted by smoke and gases for centuries. London has long been called "The Big Smoke," because it suffered regularly from smog. In December 1952 a thick yellow-brown smog hung in the air for five dark days. Daylight never appeared. It was the worst smog ever recorded, with visibility sometimes less than 20 in. (50 cm). The cold damp air full of coal smoke was trapped by a layer of warmer air above, and could not move. The air was full of soot, tar particles, and gases such as sulfur dioxide. Sulfur dioxide and water make sulfuric acid, which irritates the lungs and makes breathing difficult.

## New laws to prevent smog

About four thousand people were, in effect, choked to death. As a result laws were passed to clean up the air in London and other cities, and the effect was dramatic. Today, however, the major cause of smog is exhaust fumes from traffic. In Los Angeles, cigarette smoke and even barbecues have also been found to contribute to smog.

► **Increasing numbers of children need to inhale medication because they are suffering from asthma.**

▲ This asthmatic baby from Kiev in the Ukraine is being treated with a spray machine to ease her breathing.

## Smog and sickness

Children in particular can suffer badly from photochemical smog because it contains ozone, an unusual form of oxygen that makes children suffer from coughs, wheezing, and asthma. Childhood asthma has increased in many countries in the last few years, and it could be that vehicle pollution and related smog are to blame.

Adults also suffer from smog. In Mexico City, about half the population is suffering from some form of sickness from the city smog.

## Dangerous emissions

Vehicle exhausts are not the only cause of photochemical smog. Emissions from factories and oil, coal and gas-burning power plants also contain the gases that cause acid rain and photochemical smog. The very smallest particles in smog—under 40,000th of an inch (100th of a millimeter) across—are the most dangerous, because they get deepest into the lungs. Recent research in Eastern Europe, where there are many dirty, old-fashioned industrial factories, found that smog can even affect unborn babies in the womb.

Nuclear power plants may not be popular with some, but at the least they do not pollute the air.

# THE MYSTERY OF FOG

▲This is a poster advertising a horror film called *The Fog*.
The film uses the eerie quality of fog for maximum effect.

Fog and mist have long been used in stories to conjure up an atmosphere of mystery, suspense, danger, or doom. Filmmakers and playwrights use it, too. When a scene opens with swirls of fog around, you know immediately that something odd, terrible, or frightening is going to happen.

## FOG ON STAGE
★ Fog is even recreated on stage or on a film set to add to the atmosphere of a production. This is done with carbon dioxide that has been frozen solid, called dry ice. When hot water is poured on it, it instantly turns into a cloud of white gas. It is very chilly for the actors to work in!

# A chase in the fog

"It's moving towards us, Watson... the one thing on earth which could have disarranged my plans.... Our success and even his life may depend upon his coming out before the fog is over the path." These are the words of the famous fictional English detective, Sherlock Holmes, in one of the best-known adventures, *The Hound of the Baskervilles*. Holmes must track down a dangerous hound that lurks on the moor, but his job is made almost impossible by the fog. The fog Holmes describes is likely to be radiation fog. This is because, in the book, the moor is often thick with fog at dawn or dusk. Also, the moor is low-lying, unsheltered, and boggy, so the damp air will cool and condense quickly. Holmes also describes the fog as patchy, and this is another characeristic of radiation fog.

▲ Sherlock Holmes and Dr. Watson in a 1940s movie of the story *The Hound of the Baskervilles*.

◄ The spookiest place of all must surely be a graveyard in the fog. This scene is from a horror film called *The Terror*.

# THE FUTURE

It is becoming increasingly clear that the activities of humans, at home, and at work are slowly changing the climate. Life exists on Earth because of a layer of gases in the atmosphere that allows the sun's heat through, but stops it radiating back into space. Because these gases act rather like the glass in a greenhouse, this is known as the Greenhouse Effect. Without the Greenhouse Effect temperatures would fall below freezing.

## Global Warming

Human activity, especially since the Industrial Revolution, has poured millions of tons of extra carbon dioxide and other greenhouse gases into the atmosphere, trapping even more heat. As a result, the global climate is warming up, a problem known as "Global Warming."

▼ The earth's atmosphere acts like the roof of a greenhouse, letting heat in and keeping it there.

Of the heat coming from the sun, about 25 percent gets caught in the atmosphere, 25 percent is reflected back into space, and the rest reaches the surface of the earth.

Some heat trapped in the atmosphere returns to the earth's surface, only to be released into the air once more.

Of the heat rising from the land and oceans, as much as 50 percent may be trapped by the atmosphere. Clouds trap far more of the heat than clear skies.

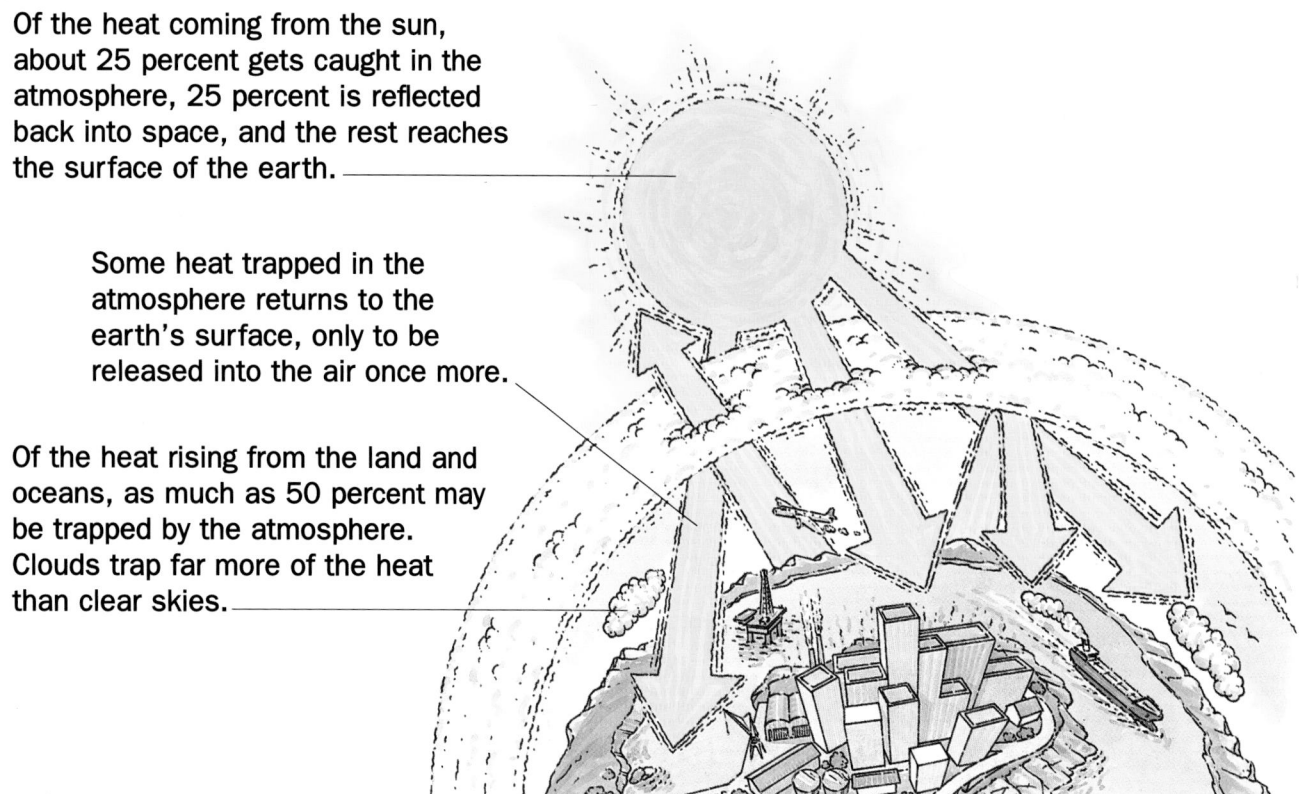

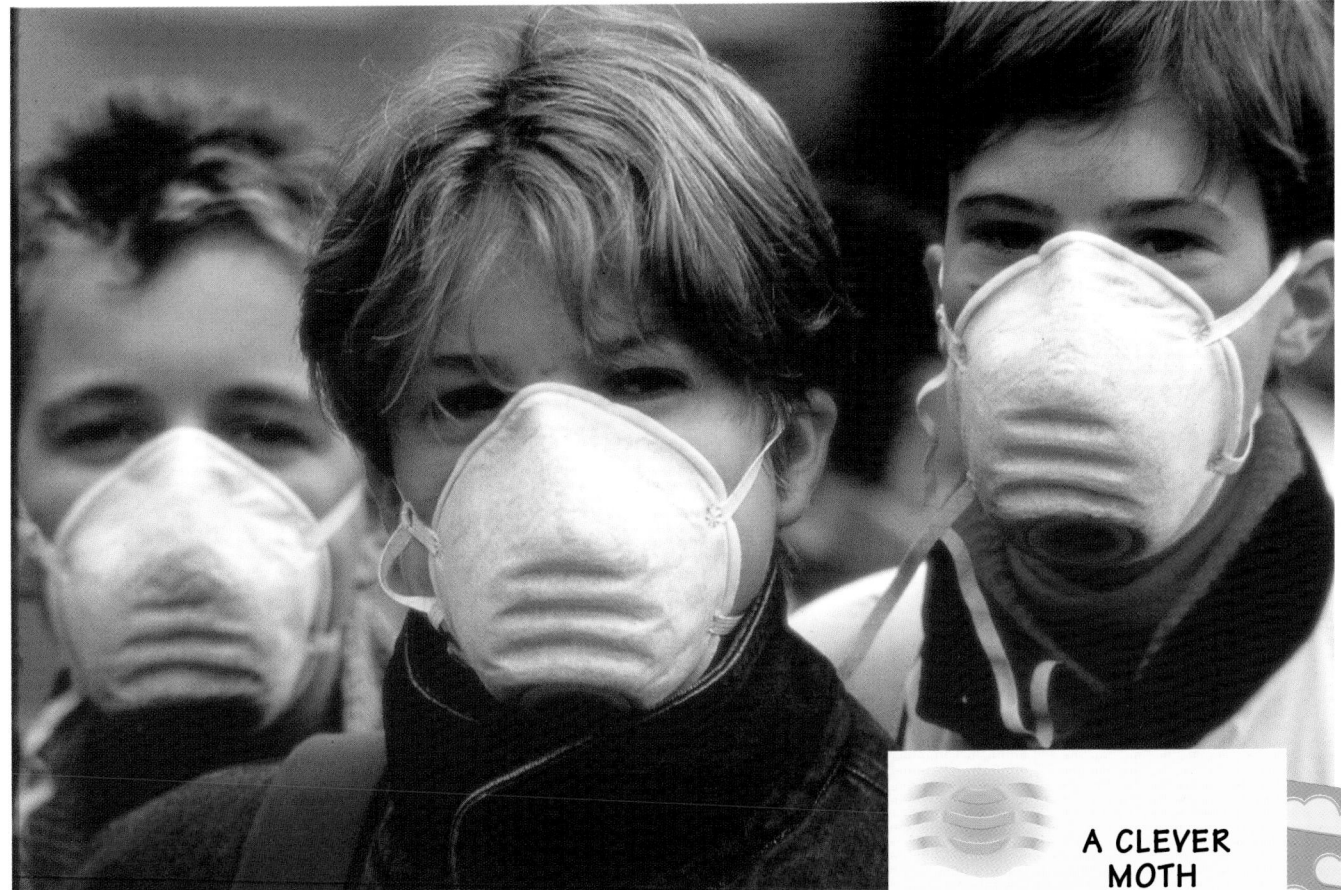

▲ Will the future for all children be that they must wear masks when air quality is poor, like these schoolchildren in the Czech Republic?

## Changing climates

Already the weather patterns are changing. Extremes of storm and flood, heat and drought, seem to be more common. Ocean currents such as the Gulf Stream, which flows from the Caribbean to northern Europe, may change direction. The climate in some places will become warmer; in others it will become colder and wetter. This means areas where mist and fog are common may move. Northern cities such as Stockholm, Copenhagen, or Montreal may become sunnier. If so, they may begin to suffer from photochemical smogs.

### A CLEVER MOTH

★ Animals can adapt well to change—even changes for the worse. In Great Britain, before the Industrial Revolution and factory pollution, the peppermoth was a silver-gray color that enabled it to hide on trees and stones. But as soot turned the landscape darker, the peppermoth turned darker, too.

# A CLEARER FUTURE

Is it possible to prevent mist and fog, or at least to stop them from being a hazard? Is it possible to prevent smog from forming? Technology has come to the help of humans before, could it do so again?

## Avoid the problem

Nobody really wants to change the weather altogether. Weather is one of the things that makes the planet interesting. It might be possible to prevent fog from forming in certain places at certain times, but it would be unrealistically expensive. Mist and fog can be blown or burned away from relatively small areas—such as airport runways—but clearing them from long roads is not practical. It would be better not to build roads in areas where fog is likely, such as in river valleys.

## Stop the smog

Smog, however, is created largely by people, so it can be reduced. The authorities in California, where cities such as San Diego and Los Angeles are badly affected by smogs, are trying to find solutions. They have decided that by the late 1990s, a small proportion of all cars must be non-polluting.

▼ Cleaner, electrically powered cars are being developed to replace cars with gasoline engines that give off dirty fumes.

## Electric cars

At the moment, the best way to cut down on motor pollution is to use cars that are fueled by electricity instead of by gasoline or diesel. But many people do not like the idea of driving the smaller, slower electric cars. Only recently have the first good electric cars become available, but they still do not have the power of gas guzzlers and are unsuitable for long journeys. Electric cars are likely to be used more and more for short trips to school or the shops.

## Save energy

What can you do to help reduce pollution and the smog? Using energy more wisely, wasting less, and recycling materials all help reduce the amount of energy that power plants have to generate, and the pollution they and factories produce making energy and new products.

◄ **Recycling the packaging of the products and foods you use every day saves on raw materials such as metals, wood, and glass.**

# GLOSSARY

**Advection** The movement of air in a sideways direction. Advection fog is produced by warm air moving across cold ground.

**Atmosphere** The thick layer of gases that surrounds the earth.

**Carbon dioxide** The gas that is breathed out by animals, and which plants "breathe" in. It is a natural part of air.

**Condense** To change from a gas or vapor into a liquid.

**Convection** The movement of air in an upward direction. Warm air rises, carrying its heat with it.

**GPS** The Global Positioning System, developed by the U.S. Army. A handheld receiver can pinpoint its position on the ground, using signals from a network of satellites in space.

**Gyro-compass** A compass that works even when tilted.

**Inversion** Turned upside down.

**Meteorologist** Someone who studies weather and climate.

**Moisture** A dampness either in the air or on a surface.

**Photochemical** A chemical reaction that is triggered by light.

**Radar** A device that uses the echoes of radio waves bouncing off objects to work out where the objects are.

**Radiation** Rays of energy, such as heat or light, coming from a source.

**Supercooled** A liquid that has cooled below its normal freezing point.

**Temperate climate** A temperate climate is a moderate climate, neither very hot, nor very cold.

**Tropical climate** A tropical climate occurs in the regions just north and south of the equator, where it is very hot.

**Vapor** The gas formed from a liquid, by evaporation.

**Visibility** The clearness of the atmosphere, which affects your ability to see.

# BOOKS TO READ

Ardley, Neil. *The Science Book of Weather*. San Diego: Harcourt Brace, 1992.

Bower, Miranda. *Experiments with Weather*. Minneapolis, MN: Lerner Publications, 1993.

Casey, Denise. *Weather Everywhere*. New York: Simon & Schuster Children's Books, 1994.

Cosgrove, Brian. *Weather* (Eyewitness). New York: Knopf Books for Young Readers, 1994.

Flint, David. *The World's Weather*. Austin, TX: Thomson Learning, 1993.

Gardner, Robert & Webster, David. *Science Projects About Weather* (Science Projects). Springfield, NJ: Enslow Publishers, 1994.

Kerrod, Robin. *The Weather* (Let's Investigate Science). Tarrytown, NY: Marshall Cavendish, 1994.

Mason, John. *Weather and Climate* (Our World). Parsippany, NJ: Silver Burdett Press, 1991.

Mogil, H. Michael & Levine, Barbara G. *The Amateur Meteorologist: Explorations and Investigations* (Amateur Science). Danbury, CT: Franklin Watts, 1993.

Pollard, Michael. *Air, Water and Weather* (Discovering Science). New York: Facts on File, 1987.

# INDEX

© Copyright 1997 Wayland (Publishers) Ltd.

PHOTO ACKNOWLEDGMENTS
Pages 4: Trip, 5/6: Zefa Picture Library, 7b: Panos Pictures; tr: Trip; 9: Hutchinson Library, 10/11: Panos Pictures, 13: Zefa Photo Library, 15: Trip, 16: Frank Lane Phot Agency; 18/19: Trip; 21: Panos Pictures, 21t; Zefa Phot Library; cr; Frank Lane Photo Agency, 22: Trip, 23: Zefa Photo Library, 25: Robert Harding Associates, 26: Zefa Picture Library, 27: Frank Spooner Pictures, 28: Frank Lane Phot Agency, 29/31: Trip, 32: Robert Harding Associates, 33: Hutchinson Library, 34: Tony Stone Associates, 35: Trip, 38: Still Pictures, 39: Trip, 40: Ronald Grant Archive, 43: Frank Spooner Picture Agency, 44: Peugeot.